Relativity of Time and Energy()orbit

Thank you very much to CEO & Staff of Amazon kdp.

The expansion of:
WORLD OF CHITTA

T. Adverb

Introduction

Publication in this article. It is an expansion of some parts from the books "World of Chitta". In terms of Orbit in physics, for energy development and understanding of the true dimension of time, in both events and places without limitation in common sense understanding.

T. Adverb

Reference book:

World of Chitta V.1 & V.2 .... By T. Adverb

SUMMARY OF CONTENT ON KNOWLEDGE BASE:

PHYSICS

1. All things and events are ENERGY

2. The infinite form of Energy in a state, where the Time stands still

3. Reality in the Laws of Physics and Quantum, that is the same piece

MATTER AND ELEMENTS

The Origins of Matter and Elements UNDERSTANDING. Including the decay of matter and radioactive transformation. That can be used for Creating new elements to suit tasks and even.

THE DIMENSION OF TIME

1. The Dimension of time for unequal time in each event, each case and each place ....

2. In order to gain an understanding of the temporal dimension of Orbit, that affects movement, distance and speed.

3. To understanding: reflections at some scale in natural phenomena

Relativity of Time and Energy()orbit

The reasons for pressing physics are not true in every case()null. That comes from putting too much emphasis on speed. So, making it impossible to transcend the SPEED OF LIGHT. Quantum physics is supposed to be largely imagination in its interpretation. The further it moves away from the true knowledge....

True natural law, made up of Orbits with inverse Time and Energy.

ORBIT()looper

An Orbit is the Cycle of processes that give rise to the makeup of matter and things. Orbit is a constant coefficient. The specific components of each Orbit have processes that determine its characteristics. Each element has its own specific Orbital value. The specific orbits of many elements Orbit made up the great Orbit of all things.

All orbital elements have time and energy that are inversely constant. Orbits with more elements complete their cycles more slowly than orbits with fewer elements.

For example: The Sun's orbit be looper of Hydrogen as its main component. The Sun's internal orbit has very short cycles. The structure of the Sun therefore has a lot of energy.

All matter decays over time. Metal elements decompose slowly, because the time of looper processing is slower than liquid and gas.

And with the inverse relational of time and energy. Which occurs within the orbital elements. Can be created into a mathematical formula that is true in every case of matter, time and energy in all fields of physics. From which I have made the following preliminary inferences:

1. The Orbit of all things is made up of the orbits of many substances and elements coming together.

$$Ob_t = Ob_1 + Ob_2 + ... Ob_n$$

Ob: Orbit Coefficient

Ob_t: total Orbital Coefficient

Ob_n: Orbital Coefficient Sequence

2. Orbit is the value of the movement performed on a cycle()π. The value is the inverse of Energy and Time in Orbit. And the Time be not negative value.

$$Ob.\pi = E / \sqrt{T_{ob}}$$

and then, $T_{ob} = 0 \; ; \; E = \infty$

T_{bo}: Time in Orbital Coefficient

3. Energy is therefore the orbital coefficient multiplied by the orbital coefficient's time.

$$E = Ob.\pi \sqrt{T_{ob}}$$

"Time is therefore a one of the forms of energy"

In every event and place()true: therefore, it can be realized by understanding the Coefficients of Orbit...

It is possible to achieve speeds above the **SPEED OF LIGHT**. By understanding the Orbital system.

In additional: We can also understand the actual system of time and space, on a commonsense level.

Example: Suppose we were born on the planet Sun. And we have a body()orbit, that has been conditioned. When we go to the World. We will see people and everything in the World including thoughts and movement.... Everything went very slowly. And we will instantly become **the Superman**.

The Path of Orbit

Definition 1:

"There is no beeline travel in all normal Orbits"

Ob ≠ Line

The system of things is Orbital. For this reason, all matter in the path of trajectories has a valve of "PI()π" element. This is necessary in the calculation of every path equation. And the journey that is true throughout this universe (except for systems that do not have curved orbits.

For example:

1. We all move on the spherical surface of the earth

2. Two points in a fall event are vertical downward movement on radial movement, that move away from the same point all the time

3. The actual location of the intended launch object into space. It is a point that aways moves from its current position on the basis of time

Definition 2:

"Geometric dimensions determine the shape of time"

$$T_{\underline{ob}} = \lambda Ob$$

λOb: Orbital geometric shape

The time determines energy. And at the same time, **Time is Energy**. Therefore, the movement is in a straight line. So, this can be made possible by creating an orbit that has an inferred shape()λOb.

For example: Snowflakes, Diamonds, prisms, etc.

Thks. 4 Ur supported the books: **World of Chitta**

T. Adverb

Sincerely hope 2 see U again ….

Knowledge base's Following in "World of Chitta" the complete edition

Manuscript()Thai:

บทนำ

สิ่งตีพิมพ์นี้เป็นบทความ ของการขยายความบางส่วน จากหนังสือ **World Of Chitta** ในส่วนของการนำเรื่องโคจรไปใช้งาน ในทางฟิสิกส์ เพื่อการพัฒนาทางพลังงาน และความเข้าใจในมิติแห่งเวลาที่เป็นจริง ในทั้งเหตุการณ์ และสถานที่ โดยไม่มีข้อจำกัดในการทำความเข้าใจ ทางสามัญสำนึก

พุฒพส ตระกูลทอง

Reference book:

World of Chitta .... T. Adverb

2

สาระของเนื้อหา:

1.ความเข้าใจเรื่องกำเนิดของสะสารและธาตุ รวมถึงการสูญสลายของสะสาร และการแปรรูปทางกัมมันตรังสี

2.การสร้างธาตุใหม่ เพื่อให้เหมาะสมกับงาน และเหตุการณ์

3.ความเข้าใจเรื่องเวลาที่ไม่เท่ากันในแต่ละสถานะการณ์ ในแต่ละสถานที่ แม้ในสภาวะของการหลับฝัน

4.ความเป็นจริงในกฎทางฟิสิกส์ และควอนตั้ม

5.การเข้าใจในรูปแบบแห่ง อนันต์ของพลังงาน ในสถาวะเวลาที่หยุดนิ่ง

3

สัมพันธภาพแห่งเวลาและพลังงาน( )โคจร

เหตุที่กฎทางฟิสิกส์ยังไม่เป็นจริงในทุกกรณี( )null นั้นเกิดมาจาก การให้ความสำคัญกับความเร็วมากเกินไป ทำให้ความเร็วแสงไม่สามารถถูกก้าวข้ามไปได้ ควอนตั้มฟิสิกส์ที่ควรจะเป็นส่วนใหญ่จะใช้จินตนาการไปตีความ ยิ่งทำให้ออกห่างจากองค์ความรู้ที่แท้จริงมากขึ้นไปอีก...

กฎธรรมชาติที่เป็นจริง( )true ประกอบจากโคจร โดยมีเวลาและพลังงานเป็นปฏิภาคกัน

โคจร

โคจรคือรอบของขบวนการ ซึ่งทำให้เกิดการประกอบขึ้นเป็นสะสาร และสรรพสิ่ง โคจรมีค่าคงที่ องค์ประกอบจำเพาะของแต่ละโคจร มีขบวนการเป็นตัวกำหนดคุณลักษณะ ในธาตุแต่ละชนิดจะมีค่าจำเพาะทางโคจรเป็นของตนเอง โคจรจำเพาะของธาตุหลายๆ โคจร ประกอบกันขึ้นเป็นโคจรใหญ่ เป็นสรรพสิ่งๆ

องค์ประกอบของโคจรทั้งหลาย มีเวลาและพลังงานเป็นค่าที่ผกผันกัน ในรอบของโคจรที่มีองค์ประกอบมาก จะครบรอบได้ช้ากว่าโคจรที่มีองค์ประกอบน้อย

ตัวอย่าง: พระอาทิตย์มีโคจรไฮโดรเจนเป็นองค์ประกอบหลัก โคจรภายในของพระอาทิตย์ซึ่งมีรอบของขบวนการที่สั้นมาก โครงสร้างของพระอาทิตย์จึงมีพลังงานมาก เป็นต้น

4

สะสารทั้งหลายย่อมมีการสลายตัวไปตามกาล ธาตุโลหะมีการสลายตัวได้ช้า เพราะเวลาในรอบของขบวนการมีช้ากว่าสะสารที่เป็นของเหลว และก๊าซ

และด้วยความสัมพันธ์ผกผันกันของเวลากับพลังงาน ซึ่งเกิดขึ้นภายในองค์ประกอบของโคจร สามารถนำมาสร้างเป็นสูตรทางคณิตศาสตร์ที่เป็นจริง()true ในทุกๆกรณีของสะสาร เวลา และพลังงานทางฟิสิกส์ในทุกสาขาได้ ซึ่งผู้เขียนอนุมานเบื้องต้นไว้ดังนี้:

1.โคจรของสรรพสิ่ง เกิดจากโคจรของหลายๆ สะสารและธาตุโคจร... รวมกันเป็นโคจร

$$Ob_t = Ob_1 + Ob_2 + ... Ob_n$$

Ob: สัมประสิทธิ์โคจร

Ob_t: สัมประสิทธิ์โคจรรวม

Ob_n: ลำดับสัมประสิทธิ์โคจร

2.โคจรเป็นค่าของขบวนการที่กระทำวนเป็นรอบรัศมี()π โดยมีค่าเป็นการผกผันของพลังงานกับเวลาในโคจร และเวลาไม่มีค่าเป็นลบ

$$Ob.\pi = E / \sqrt{T_{ob}}$$

และเมื่อ $T_{ob} = 0$; $E = \infty$

T_{ob}: เวลาในสัมประสิทธิ์โคจร

3.ดังนั้นพลังงานจึงเป็นค่าสัมประสิทธิ์ของโคจรนั้นๆ คูณด้วยค่าของเวลาในโคจร

$$E = Ob.\pi \sqrt{T_{ob}}$$

"เวลาเป็นหนึ่งในค่าของพลังงาน"

ในทุกๆ กรณี ที่เกี่ยวข้องกับเหตุการณ์ และสถานที่จึงเป็นจริง() true ด้วยการเข้าใจในสัมประสิทธิ์ของโคจร ที่เกิดจากพันธะระหว่าง เวลา()T_{ob} กับพลังงาน()E

นอกจากนั้น: ยังสามารถทำความเข้าใจในระบบเวลา และสะสารที่เป็นจริง ของทุกๆ กรณี ทุกๆ สถานที่ของจักวาล ในระดับสามัญสำนึกได้...

ตัวอย่าง: สมมุติถ้าเรามีกำเนิดอยู่บนดาวพระอาทิตย์ มีร่างกาย()โคจร ที่ถูกปรับสภาพแล้ว เมื่อเรามายังโลกมนุษย์ เราจะเห็นผู้คนและทุกสิ่งบนโลกทั้งเรื่องความนึกคิด และการเคลื่อนไหวที่ช้ามาก เราจะกลายเป็นซุปเปอร์แมนของโลกไปทันที...

ขอขอบคุณที่ช่วยสนับสนุนหนังสือ World of Chitta

พุฒพส

เส้นทางแห่งการโคจร

ขอแสดงความขอบคุณมายัง อเมซอน เคดีพี ไว้เป็นอย่างสูง

<u>การขยายความเรื่อง:</u>
สัมพันธภาพแห่งเวลาและพลังงาน()โคจร

T. Adverb

2

บทนำ

สิ่งตีพิมพ์นี้เป็นบทความ ของการขยายความเรื่อง
สัมพันธภาพแห่งเวลาและพลังงาน()โคจร

จากหนังสือ **World Of Chitta** ในส่วนของการนำเรื่องโคจรไปใช้งาน ในทางฟิสิกส์ เพื่อการพัฒนาทางพลังงาน และความเข้าใจในมิติแห่งเวลาที่เป็นจริง ในทั้งเหตุการณ์ และสถานที่ โดยไม่มีข้อจำกัดในการทำความเข้าใจ ทางสามัญสำนึก

พุฒพส ตระกูลทอง

Reference book:

Relationship of the Time and Energy()orbit

.... T. Adverb

World of Chitta .... T. Adverb

3

สาระของเนื้อหา:

1. เพื่อให้เกิดความเข้าใจในมิติทางเวลาของโคจร ที่มีผลต่อการเคลื่อนที่, ระยะทาง และความเร็ว

2. เข้าใจความเป็นจริงทางฟิสิกส์ และควอนตั้ม

3. ภาพสะท้อนบ้างขณะ ในปรากฏการทางธรรมชาติ จากความหนาแน่นของบรรยากาศ

4

เส้นทางแห่งการโคจร

นิยามที่ 1:

"การเดินทางที่เป็นเส้นตรงไม่มีในโคจรทั้งหลาย"

$$Ob \neq Line$$

ระบบของสรรพสิ่งเป็นโคจร ด้วยเหตุนี้ สะสารทั้งหลายซึ่งอยู่ในเส้นทางแห่งโคจรจึงมีค่า "พาย()π" ซึ่งเป็นสิ่งจำเป็นในทางการคำนวณในทุกๆ สมการของเส้นทาง และการเดินทางที่เป็นจริงในทั่วทั้งเอกภพนี้(ยกเว้นกับเอกภพที่ไม่มีระบบโคจรเป็นส่วนโค้ง)

ตัวอย่าง:

1. เราทั้งหลายล้วนเคลื่อนที่ไปบนพื้นผิวโลกที่เป็นทรงกลม
2. จุด 2 จุดในเหตุการณ์การตกจากที่สูง เป็นการเคลื่อนลงในแนวดิ่ง บนการเคลื่อนในวงรัศมี ที่เคลื่อนออกจากจุดเดิมตลอดเวลา
3. ตำแหน่งที่แท้จริงในเป้าหมาย ของการส่งวัตถุออกไปในอวกาศ เป็นจุดที่เคลื่อนออกจากตำแหน่งปัจจุบันเสมอบนฐานแห่งเวลา()พลังงาน

นิยามที่ 2:

"มิติในรูปทรงทางเรขาคณิตเป็นตัวกำหนดสัณฐานแห่งเวลา"

$$T_{ob} = \lambda Ob$$

λOb: รูปทางเรขาคณิตของโคจร

เมื่อเวลาเป็นตัวกำหนดพลังงาน และในขณะเดียวกันเวลาก็คือพลังงาน ดังนั้นการเคลื่อนที่ที่เป็นเส้นตรง จึงทำให้เกิดขึ้นได้ด้วยการสร้างโคจรที่เป็นรูปทรงตามอนุมาน()λOb

ตัวอย่าง: เกร็ดหิมะ เพชร ปริซึม เป็นต้น

ขอขอบคุณที่ช่วยสนับสนุนหนังสือ **World of Chitta**

พุฒพส

MANUSCRIPT (bule print):

Introduction

Publication in this article. It is an expansion of some parts from the books "World of Chitta". In terms of Orbit in physics, for energy development and understanding of the true dimension of time, in both events and places without limitation in common sense understanding.

T. Adverb

Reference book:

World of Chitta V.1 & V.2 .... By T. Adverb

Relationship of the Time and Energy()orbit

The reasons for pressing physics are not true in every case()null. That comes from putting too much emphasis on speed. So, making it impossible to transcend the SPEED OF LIGHT. Quantum physics is supposed to be largely imagination in its interpretation. The further it moves away from the true knowledge....

True natural law, made up of Orbits with inverse Time and Energy.

ORBIT()looper

An Orbit is the Cycle of processes that give rise to the makeup of matter and things. Orbit is a constant coefficient. The specific components of each Orbit have processes that determine its characteristics. Each element has its own specific Orbital value. The specific orbits of many elements Orbit made up the great Orbit of all things.

All orbital elements have time and energy that are inversely constant. Orbits with more elements complete their cycles more slowly than orbits with fewer elements.

Summary of Content:

1. The Origins of Matter and Elements UNDERSTANDING. Including the decay of matter and radioactive transformation

2. Creating new elements to suit tasks and even

3. UNDERSTANDING of unequal time in each event, in each place

4. Reality in the Laws of Physics and Quantum

5. The infinite form of Energy in a state, where the Time stands still UNDERSTANDING

For example: The Sun's orbit be looper of Hydrogen as its main component. The Sun's internal orbit has very short cycles. The structure of the Sun therefore has a lot of energy.

All matter decays over time. Metal elements decompose slowly, because the time of looper processing is slower than liquid and gas.

And with the inverse relational of time and energy. Which occurs within the orbital elements. Can be created into a mathematical formula that is true in every case of matter, time and energy in all fields of physics. From which I have made the following preliminary inferences:

1. The Orbit of all things is made up of the orbits of many substances and elements coming together.

$$Ob_t = Ob_1 + Ob_2 + ... Ob_n$$

Ob: Orbit Coefficient

Ob_t: total Orbital Coefficient

Ob_n: Orbital Coefficient Sequence

2. Orbit is the value of the movement performed on a cycle()π. The value is the inverse of Energy and Time in Orbit. And the Time be not negative value.

$$Ob.\pi = E / \sqrt{T_{ob}}$$

and then, $T_{ob} = 0 \, ; \, E = \infty$

T_{bo}: Time in Orbital Coefficient

3. Energy is therefore the orbital coefficient multiplied by the orbital coefficient's time.

$$E = Ob.\pi \sqrt{T_{ob}}$$

"Time is therefore a one of the forms of energy"

In every event and place()true: therefore, it can be realized by understanding the Coefficients of Orbit...

It is possible to achieve speeds above the **SPEED OF LIGHT**. By understanding the Orbital system.

In additional: We can also understand the actual system of time and space, on a commonsense level.

Example: Suppose we were born on the planet Sun. And we have a body( )orbit, that has been conditioned. When we go to the World. We will see people and everything in the World including thoughts and movement.... Everything went very slowly. And we will instantly become **the Superman.**

Thks. 4 Ur supported the books: **World of Chitta**

T. Adverb

2

Introduction

Publication in this article. It is an expansion of
Relationship of the Time and Energy()orbit

That are some parts from the books "World of Chitta". In terms of Orbit in physics, for energy development and understanding of the true dimension of time, in both events and places without limitation in common sense understanding.

T. Adverb

Reference book:

Relationship of the Time and Energy()orbit

……… T. Adverb

World of Chitta V.1 & V.2   ………………. By T. Adverb

3

Description of Content:

1. In order to gain an understanding of the temporal dimension of Orbit, that affects movement, distance and speed.

2. For understand the reality of Physics and Quantum

3. To understanding: reflections at some scale in natural phenomena

The Path of Orbit

Definition 1:

"There is no beeline travel in all normal Orbits"

$$Ob \neq Line$$

The system of things is Orbital. For this reason, all matter in the path of trajectories has a valve of "Pl()π" element. This is necessary in the calculation of every path equation. And the journey that is true throughout this universe (except for systems that do not have curved orbits.

For example:

1. We all move on the spherical surface of the earth
2. 2 points in a fall event are vertical downward movement on radial movement, that move away from the same point all the time
3. The actual location of the intended launch object into space. It is a point that aways moves from its current position on the basis of time

Definition 1:

"Geometric dimensions determine the shape of time"

$$T\underline{ob} = \lambda Ob$$

λOb: Orbital geometric shape

The time determines energy. And at the same time, **Time is Energy**. Therefore, the movement is in a straight line. So, this can be made possible by creating an orbit that has an inferred shape()λOb.

For example: Snowflakes, Diamonds, prisms, etc.

Thks. 4 Ur supported the books: **World of Chitta**

T. Adverb

Sincerely hope 2 see U again ....

Knowledge base's Following in "World of Chitta" the complete edition

www.ingramcontent.com/pod-product-compliance
Lightning Source LLC
Chambersburg PA
CBHW072057230526
45479CB00010B/1123